Smart Picture Frame

Raspberry Pi DIY projects

CHRISTOPHER DOCK

ISBN: 978-1-952930-09-6

North American Offices
Christopher Dock
3533 Rum River Drive
Anoka, MN 55303
USA

European Offices
Christopher Dock
Ludwig-Richter-Straße 15
60433 Frankfurt
Germany

Limits of Liability and Disclaimer of warranty

We work hard to provide the most accurate research on options for self-publishing books. All information provided in Smart

Picture Frame Raspberry Pi DIY projects has been provided in good faith, however, we make no representations, warranty or guarantees of any kind, express or implied, that this information will be complete, accurate, adequate, error free or will not change in the future. Information in this book may not constitute the most up-to-date information available. Smart Picture Frame Raspberry Pi DIY projects is for general informational purposes only and should not be relied upon in making legal or financial decisions.

To the maximum extent permitted by law, we disclaim all warranties, whether express or implied, including the implied warranties of merchantability and fitness for a particular purpose. To the maximum extent permitted by law, we disclaim any and all liability in the event any information, commentary, analysis, or opinions contained in this book prove to be inaccurate, incomplete or unreliable, and under no circumstances will we be liable to you for any lost profits, revenues, information, or data or incidental damages arising out of or related the information provided by this book.

Information provided in this book is descriptions about electronics as well as various tools that can be used to make your own electronic creations. The tips and information is intended to be helpful but should not be considered a replacement for any manuals, documentation or training for operation of such tools.

DEDICATION

To my wife who has been supportive of all my efforts, and who has not given me grief over piles of wood, electronics parts and half complete projects all over the house. She also suffered through the occasional evening requiring complete quiet. I also cannot give enough credit to my sons who acted as my editors.

CONTENTS

1 A SMART IDEA

Deciding on what to do is typically the hardest part of getting started on a new project. Given a completely blank slate should be a feeling of freedom as there are not any limitations. I could work on a smartphone app to track the children, a cloud application for distributed expenses tracking, a program to convert between metric units to imperial units or perhaps just a build tree house for the boys.

Other than an obsession with small computers, I have always had an interest in creating a super duper smart picture frame. Each time we visit my parents we load up their computer with a new set of photos. Once this is done I set the screensaver to point to the latest batch of photos. This is not the most flexible solution but more to the point, running a full computer with a monitor 24 hours a day as a picture frame is not very friendly to either the environment or the electricity bill.

Due to the Covid 19 “situation” I have never had more free evenings at home to work on projects. In addition to all of the free time, there has never been better for to create smart devices or even create the next big IoT device. There is a plethora of small form factor devices in the shape of micro-controller boards and small computers which can be purchased to make your ideas a reality. It is just a matter of picking the correct hardware for your project.

2 PLATFORMS

Computers just keep getting faster, smaller and cheaper. Ignoring normal desktop computers for a second still leaves a lot of small form factor computers about the size of the NUC[1] or the Apple TV[2]. To be fair even these are really quite small and could easily be hidden behind a monitor or under a book, but there are other devices that are much smaller.

The most interesting computer in my opinion is the Raspberry Pi. The Raspberry Pi was first unveiled in February 2012 as a single board computer with 512MB ram, a 100MB Ethernet ports and two USB ports. The operating system is installed on a SD card and the whole solution can be powered with a cell phone charger.

1 The NUC has the dimensions of 4.6 inches (118 mm) x 4.4 inches (112 mm) x 1.55 inches (39 mm).

2 Height: 1.4 inches (35 mm), Width: 3.9 inches (98 mm), Depth: 3.9 inches (98 mm)

It is not the specifications that are world shaking but the fact this credit card sized computer can be bought for 35 US Dollars, can use most standard pc accessories and can run most of the same software packages that also run on desktops and servers. The global interest when this was released was intense, within hours the expected 10,000 units were sold out and 100,000 units were put on back order. It turned out that the Raspberry Pi Foundation had expected demand of up to ten thousand as a upper limit for the educational community bu they somehow missed the pent up demand by hobbyists and businesses.

In the last ten years since, the Raspberry Pi foundation has taken this very successful basic computer and has continued to enhance and expand its capabilities. The original Raspberry Pi has morphed into a collection of different Raspberry Pi's each with different strengths and weaknesses.

Deciding on which one of the many current Raspberry Pi's you might wish to use in a project can be challenging due to the large number of options.

Classic - Raspberry Pi

The classic is just the largest of the Raspberry Pi[3] in their family of small factor boards.

Figure 1: Raspberry Pi v4 with after market heat sinks installed

This is just a tiny computer. The most current Raspberry Pi v4 supports WiFi, Bluetooth as well as an 1GB Ethernet connection and 4 USB ports. A standard desktop computer has a hard disk to store the operating system and program files. The Pi uses a SD card as its hard disk. The operating system is Linux and the entire ecosystem of open source software is generally available. The Pi has a complete desktop environment which supports not only a web browser but also the Libre Office Suite for your word processing or spreadsheet needs.

3 The Pi is 2.22cm x 3.37cm

There are a lot of computers in the world but what differentiates the Raspberry Pi with your home desktop is the user-accessible header pins, Figure 2.

Figure 2: Raspberry Pi 4 with header pins highlighted

The header pins allow you to easily control other devices or custom circuits. It is probably the Arduino micro-controller boards that first made this style of interface popular but this access allows the user to create circuits or link external devices for measuring or interacting with the world.

<u>Specifications</u>
CPU: Quad Core Cortex-A72 ARM 64-bit SoC
Speed: 1.5GHz
Ram: 2GB, 4GB or 8GB depending on Model
IEEE 802.11ac WiFi at 2.4 GHz and 5.0 GHz
Bluetooth 5.0, and BLE
Gigabit Ethernet Port for Wired Connection

USB 2.0: - 2 Ports
USB 3.0: - 2 Ports
2 HDMI Ports for Dual Monitor Support

Stealth Platform - Raspberry Pi 400

The full Raspberry Pi is a nice small platform computer, but to be perfectly honest, it looks a bit like a beetle with the wires coming out of all sides once all of the peripherals and cables are connected to it.

Yet, a more refined solution exists. The Raspberry Pi foundation has also created the Raspberry Pi 400 which embeds a Raspberry Pi into a keyboard and routes all of the cable connections and header pins to the backside.

Figure 3: Raspberry Pi 400 keyboard

Figure 4: Back side of Raspberry Pi 400

The Raspberry Pi 400 essentially becomes an all-in-one computer once you connect it to a monitor.

It has essentially the same specifications as a normal Raspberry Pi 4 but with two slight differences. The first difference is that it is limited to 4GB[4] ram. The second and more advantageous difference is that it runs slightly faster than a regular Pi at 1.8GHz.

Embedded Platform - Raspberry Pi 2 Zero 2

The full size Raspberry Pi is a small form factor computer at 5.6cm x 8.5cm. Yet even this small physical size, when coupled by the thickness of four USB ports, can make it physically difficult to slip this computer into an equally tiny DIY project.

The Raspberry Pi Foundation created a much smaller 6.5cm x 3cm version called the Raspberry Pi Zero 2 W. This Pi has virtually the same technical specifications as its full sized brother but doesn't have the header pins[5] but is limited to a single micro USB port.

4 The Raspberry Pi 4 can be purchased in 2, 4 and 8GB models

5 It is possible to solder header pins to the board if you need them.

Figure 5: Raspberry Pi Zero 2 W

This makes for a very flat computer than can easily be embedded but still can be contacted using either WiFi or Bluetooth. It would be perfect device except it is a bit resource constrained at 1GHz top speed and 512MB ram.

The important point is that this is a computer and needs to be operated like every other computer. The operating system needs time to boot up when you turn it on but more importantly it needs to be shutdown in an orderly manner to prevent corrupting the operating system on the SD card.

Pi Compute Module

The Raspberry Pi compute module offers similar capabilities to a Raspberry Pi but has been shrunk down to a smaller form factor. The compute module replaces the familiar I/O from the standard Pi with two 100 pin mezzanine connectors.

In order to use this board it is necessary to

have a carrier board[6] to interface with it. This board can be used for personal projects but due to its size, functionality and interface it is more appropriate for industrial embedded systems.

Raspberry Pi Pico

The Raspberry Pi Pico is completely different from the rest of the Raspberry Pi family. All of the Pi's until this point have been small computers of varying size and power whereas the Pico is a micro controller based device.

The difference between the two is that a computer has an operating system while a micro-controller has a single program loaded on it[7]. A micro-controller is actually a perfect embedded solution for a lot of IoT devices. It can be turned on and off without the need to shutdown an operating system.

The downside of a micro-controller solution is the resources are quite a bit limited. The speed that they run at is the low megahertz range and memory is limited to kilobytes.

Specifications

```
Size:          21 × 51 mm
Processor:     Dual-core ARM Cortex-M0+
Clock:         Up to 133 MHz
Memory:        264 kB SRAM, 2 MB QSPI Flash

Ports:
               26x multi-function GPIO pins
```

6 Open source carrier board for the compute module. https://www.digikey.de/en/maker/projects/creating-a-raspberry-pi-compute-module-4-cm4-carrier-board-in-kicad/7812da347e5e409aa28d59ea2aaea490

7 You simple install a program on the micro-controller.

```
              3x analog inputs
              2x UARTx
              2x SPI controllers
              2x I2C controllers
              16x PWM channels

Power:        1.8-5.5 V DC
```

My project Requirements

The Raspberry Pi Foundation has provided a lot of supporting platforms any of which could be used in a DIY project.

My idea is to create my own IoT smart picture frame with a small twist. It should not only display photos but should also display the local weather forecast. Ideally, there should be a mechanism to periodically check and download any new pictures over the internet.

A smart picture frame is not much more than a small computer displaying photos onto a tiny screen. The Raspberry Pi Zero 2 W is more than enough computer for the task and it is can be connected to most screens. The Pi ecosystem also has access to a software catalog of tens of thousands of software packages which should help with implementing this solution.

Hardware and Software for this project

- Raspberry Pi Zero 2 W
- Raspberry Pi power supply
- Screen + power supply
- Video cable
- SD card
- Dropbox account
- Open Weather account
- Picture frame (optional)
- Mounting materials (optional)

3 PREPARING THE RASPBERRY PI

This chapter will discuss not only how to create our own SD card with an operating system but will also cover a lot of material that might be considered quite basic. The Raspberry Pi is an exceptional computer that is powered by Linux and as most computer users use Windows some of this material might be new. If not, just skim over this chapter about Raspberry Pi setup.

The Raspberry Pi is not a kit, simply connect your cables and peripherals and you can use it. However, unlike a standard desktop computer it is not possible to boot it up from a DVD or USB stick to install the operating system. The operating system must be installed on the SD card prepared on another computer.

The easy way

You can create your card yourself. Simply purchase a SD card and then download and install the image[8]. There are a number of different methods to do this, but actually the easiest is to download the Raspberry Pi imager[9] to a Windows computer. The program is actually pretty clever. You are not required to download an image from the Raspberrypi.org.

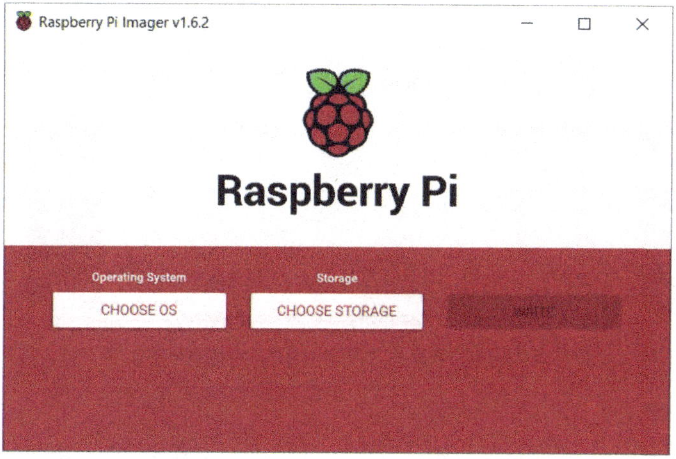

Figure 6: Raspberry Pi imager

The "choose OS" button will actually download the current version of operating system that you select the first time. The OS image will be saved in the cache if you choose to make a second SD card using that image. This is a super easy and friendly tool for creating your own SD card. The buttons are obvious, the only thing that is not obvious is the hidden setup that is available via the shortcut

8 https://www.raspberrypi.com/software/operating-systems/

9 https://www.raspberrypi.com/software/

key sequence ctrl-shift-X. This will bring up a dialog of advance options.

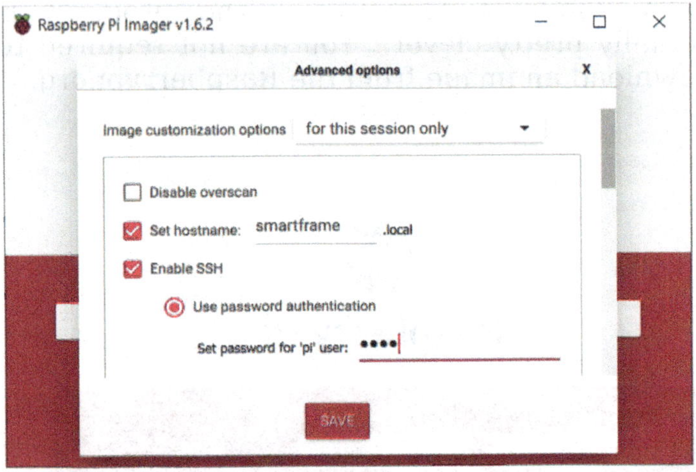

Figure 7: Modifying advanced setup options

These advanced options are pretty neat. You can change the name of the host, enable your ssh connection and set the password for the "pi" user. It is possible to set these options not only for the session but also to make them permanent defaults.

The importance of being able to enable the SSH daemon while creating the card should not be overlooked. The Raspberry Pi is just a computer and it is possible to open up a connection to it from a remote computer (ie secure session or ssh). The difficulty is that this feature is disabled by default. If it is disabled then you need to connect the Raspberry Pi with all of the peripherals as well as a monitor in order to manually enable this type of remote access. With this SSH setup enabled during the SD card creation it is then possible to simply connect to the Pi after the first boot.

This connection is only possible if the computer is connected via Ethernet to your network, otherwise you still need to boot up the Raspberry Pi and enter your WiFi connection credentials.

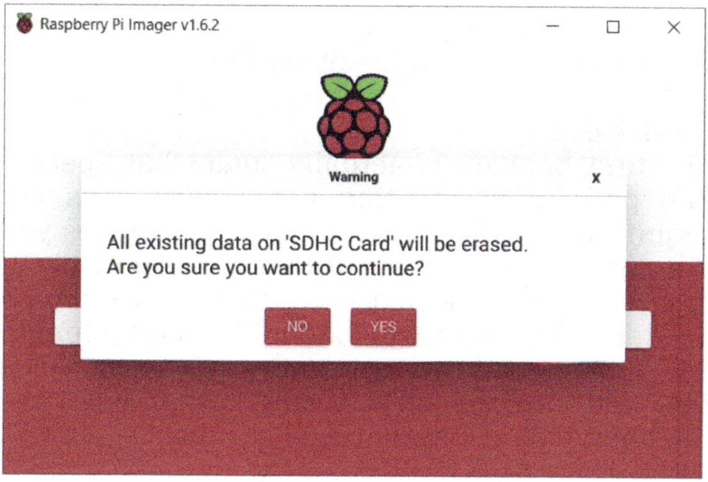

The program does give you one last prompt to ensure you do not overwrite your SD card by accident. The process of writing the SD card takes a few minutes until it is finished.

The "hard" way

Hard is relative. There is a second method of creating a SD card. Simply download your chosen operating system image and burn it to your SD card.

1. Download the image

2. Verify the sha256 hash
3. Use dd or some other tool to burn image

These steps are not too different from the easy way yet it may be more time intensive but is more manual. These tools are not native to Widows and when using Linux it is possible to make a big mess if you are not paying attention.

Don't forget that it is also usually possible to purchase a prepared SD card from the same vendor that sells the Raspberry Pi.

First Boot

The first boot up is actually somewhat special. The process of creating a SD card does not require a specific sized SD card. This gives you the option of using a card that is 8, 16 or 32GB in size. To ensure that this entire space can be used, the first boot will extend the file system to use the unused portion of the SD card. Once the file system has been expanded, the Raspberry Pi will automatically reboot.

Config

On the second boot of the Pi you will be presented with a small dialog for setting up some of the most important configurations of your Pi.

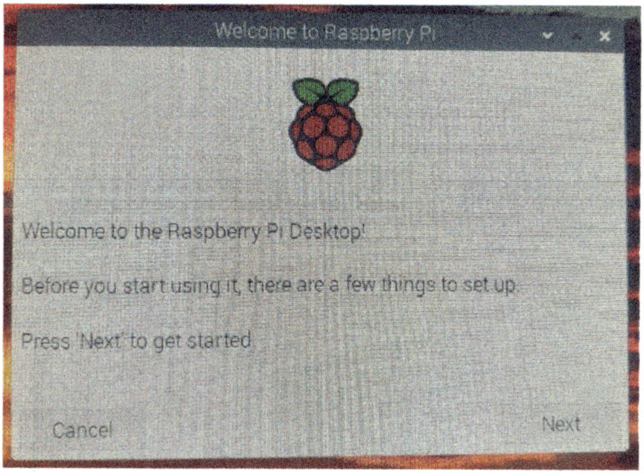

Figure 8: Dialog to guide you through setup process

All of the questions are pretty self explanatory. The most important is what language and local settings you wish to use (ie keyboard layout). Beyond that you will be offered the possibility to join a wireless network, receive some help if the display is not properly displaying, and change the password of your “pi” user.

From this point on you can continue to sit at the Raspberry Pi and work or you can connect to it with your favorite ssh client.

Old school setup

This setup can also be done from the command window using the original raspi-config program. The good news is that this will bring up a friendly text window application, but it cannot be run using standard user permissions but instead you will need to use sudo to run it.

sudo raspi-config

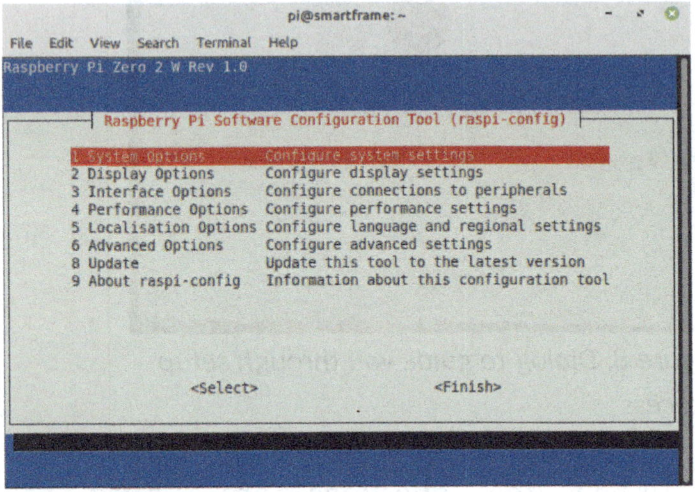

Figure 9: Old text based Raspberry Pi config tool - raspi-config

In my smartframe application I am going to be downloading files and use other files to store some tracking information. I do not know how long it would take for a brand new SD card to experience problems but as most SD cards do not have wear leveling[10] I am going to create my own solution to eliminate this problem. The solution is that my temporary files will be stored in memory. I will convert a small portion of ram into a ram disk. This is where the

10 Wear leveling is the process of spreading writing over the memory to prevent over use of a single block which would eventually lead to failure and data loss.

temporary work files will be stored.

Ram Disk

Creating a ram disk is not exceedingly difficult. There exists a configuration file, fstab[11], which is used to keep track of mounting the Linux file systems. All we need to do is to use the tmpfs filesystem which will dynamically allocate ram as a disk.

Simply add the following line to your fstab file in order to create a 15MB temporary ram based disk.

```
tmpfs /var/tmp tmpfs nodev,nosuid,size=15M 0 0
```

This ram disk can be accessed by reading or writing to the /var/tmp directory. You can make this directory as large as you want, but beware you are using your limited ram to do so. This is less of a problem on the Raspberry Pi 4 with between 2 and 8GB ram but the Raspberry Pi Zero 2 W has only 512MB of ram.

This ram disk entry can only be saved to the fstab file if you have admin privileges. This can be achieved using sudo before starting you editor.

sudo vi /etc/fstab[12]

This change will take effect after the next reboot of the Raspberry Pi.

11 The fstab file is located in the /etc directory.

12 The vi editor is not very user friendly but the Raspberry Pi has another editor "kate".

Software installation.

Installing software might be a bit different when compared to a Windows installation. The Raspberry Pi actually does have a graphical tool for installing new software called "Recommended Software". This tool only shows a very small subset of the software that can be installed.

In order to install some of your software for this project you will have to open a terminal window and type the specific install command for the package manager. These commands are fairly intuitive once you have seen them once or twice.

There is a fun little program from the Unix days called fortune. Each time it is run you will receive a random quote or joke.

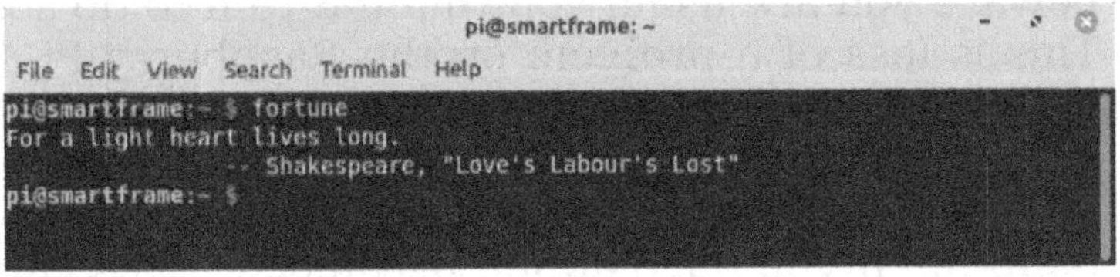

Figure 10: Sample output from fortune program

Installing this program is done with the following command.

sudo apt-get install fortune

However, before the program is installed you will be notified how much space would be required and you need to accept this by pressing the letter "Y", Figure 11.

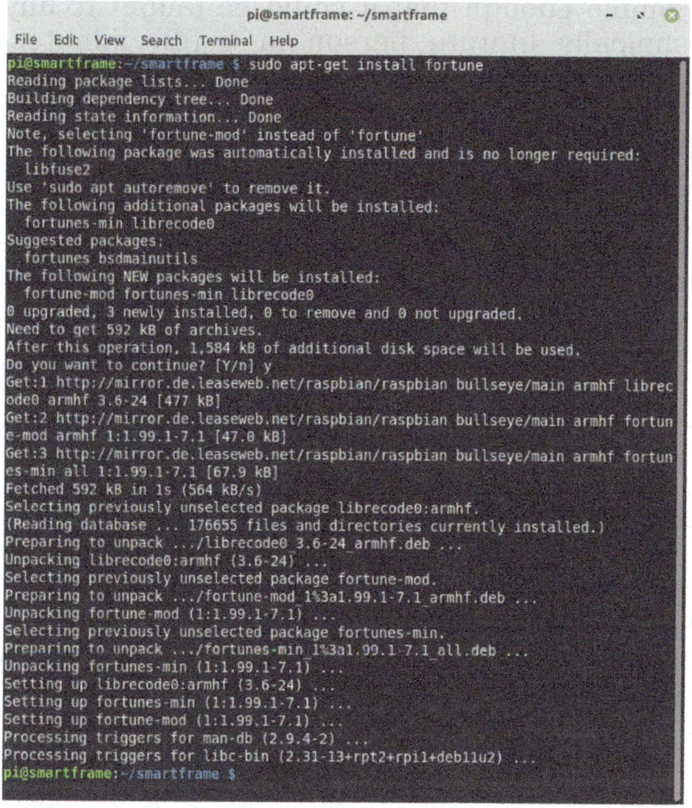

Figure 11: Example of installing software from the Raspberry Pi package repository

We will need to use this method to install some of the required software packages later.

Advanced methods

The Raspberry Pi is a very innovative area of development. The work that is being done in this area is not limited to friendly GUI programs for creating SD cards. This process is simple enough that it could be taught to any technically literate person in a few minutes. Despite how easy this is you are limited to the version of the OS on the SD card. The OS on the SD card will be initially current but over time it will get more and more out of date as updates and fixes are created.

One solution for this would be to boot the Raspberry Pi from a local network image - PXE[13]. This allows the device to receive the OS image over the network and will boot that image. This is super convenient especially if you have quite a few devices and the images for these devices often need changes. In that case, you can update a single image and make it available on the network.

This actually sounds great but can be technically rather involved creating the image and making it available on the network. The newest innovation is the ability to boot the Raspberry from such an image over the internet[14]. This is a pretty cool solution for people or libraries that might want to use the Raspberry Pi for light web browsing but don't want a lot of maintenance.

There is one final recent innovation that is somewhat related to the boot process. The standard boot device for the Raspberry Pi is an SD card but the SD card does not do wear leveling and it also does not have high

13 https://williamlam.com/2020/07/two-methods-to-network-boot-raspberry-pi-4.html

14 https://www.tomshardware.com/how-to/raspberry-pi-network-install

throughput when compared to some other USB devices. This latest change is to change the firmware so the Pi can boot from a USB device[15]. This allows USB sticks or USB external drives to be prepared with the OS image and boot from. These USB devices most likely will have faster throughput and much much more storage.

15 https://www.tomshardware.com/how-to/boot-raspberry-pi-4-usb

4 RESEARCH AND PREPARATION

Many store bought digital picture frames actually fulfill much of what we want to achieve, the special unmet requirement is the retrieval of new photographs over the internet.

Some of my development style was learned from my first boss. He was a big fan of keep it simple stupid - KISS. Trying to create some sort of connection between the picture frame over the internet direct to my house is far too complex. A better solution would be to store the photos on the internet. This would allow the picture frame to download photos and would also allow me to upload new photos.

There is no dearth of internet storage options, starting with an Amazon S3 bucket but all of the cloud providers have some similar option. Yet, if you are really on a budget you could use a free option.

Dropbox provides paid solutions but they also have a free tier which allows 2GB of data storage. This should be more than enough for our digital frame.

Dropbox

Dropbox provides a web browser interface for users to upload, download and delete stored files. This is super convenient for uploading new pictures as a person but is not helpful for a digital frame.

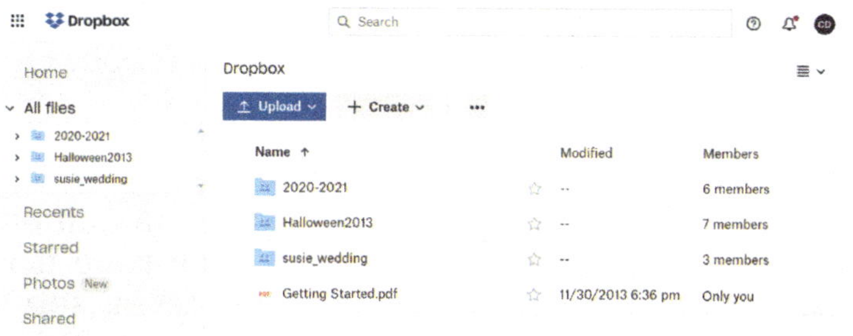

Figure 12: Dropbox web interface

A digital frame will need to operate with Dropbox without any user intervention. This can be done using the command line tool dbxcli.

This tool, dbxcli, can do the one necessary task of downloading files which is what our digital frame would need to retrieve new photographs. This tool is can run not only on Linux, but also on Windows and OpenBSD. Most importantly this program will run on the Raspberry Pi.

https://github.com/dropbox/dbxcli/releases

Once this program has been downloaded you may need to change the files permissions so it can be executed, Figure 13.

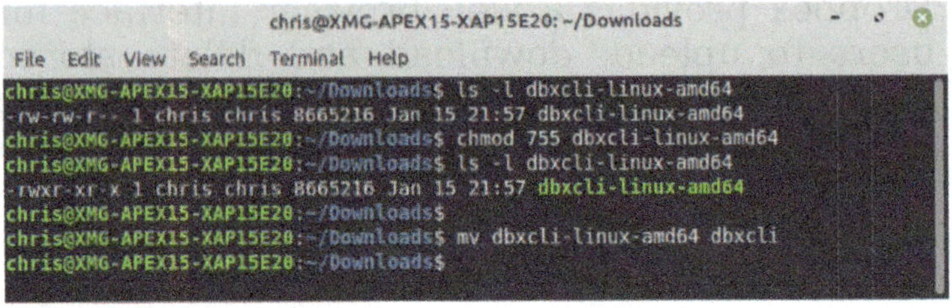

Figure 13: Changing the file permissions on Dropbox command line client

The dbxcli utility doesn't use a simple password for accessing the data each time it is used Actually a fixed "link" needs to be setup between Dropbox and this computer. What happens is the first time you run the program and attempt to do anything, such as listing the files, you are prompted with a special URL, Figure 14.

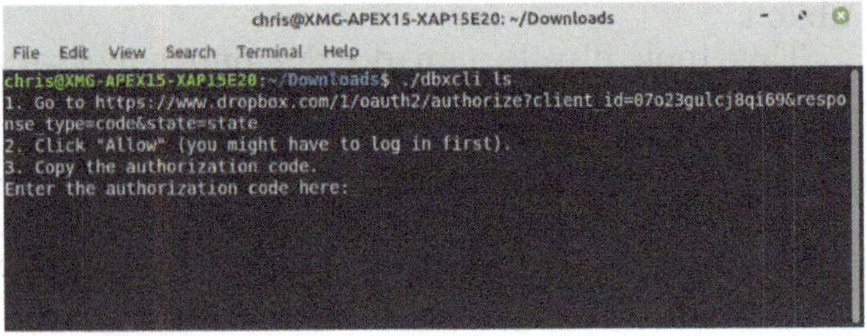

Figure 14: The first time Dropbox command line is run

This URL contains a couple of unique codes which will be used to help join this computer to the Dropbox account. Using this URL will prompt you to enter your Dropbox username and password.

When you use this URL you will be prompted with a warning, Figure 15, so you realize this program will have full access to your Dropbox account.

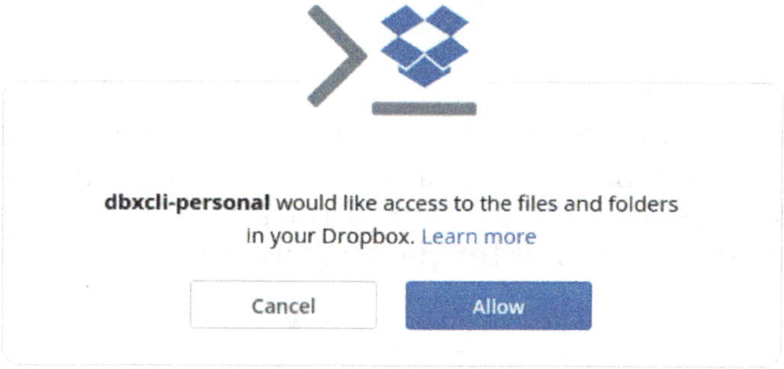

Figure 15: Message with implications of continuing

This is a great warning but as this is the behavior we are trying to achieve we can press the allow button. Once you press allow the secret passcode is displayed, Figure 16.

Figure 16: Returned passcode

This passcode needs to be entered back at the command prompt, Figure 14. Once this password has been entered, you can run the Dropbox command line program without any further need to enter passwords, Figure 17.

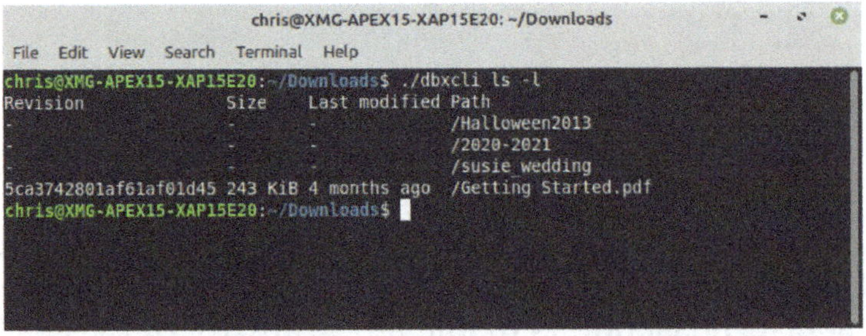

Figure 17: Sample directory listing of Dropbox account

Note: It is not necessary to know where this "link" is actually stored, but all the necessary information for the dbxcli command is stored in the ~/.config/dbxcli directory.

Dropbox supports folders as well as normal files.

You can list the files in the root of the Dropbox account, Figure 17, but you can also list the files inside of those directories Figure 18.

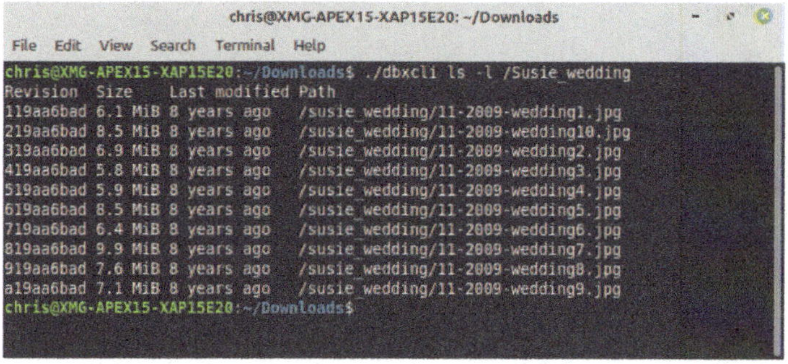

Figure 18: Directory listing of a subdirectory

There is a small quirk with this command line tool. The copy command(cp) actually refers to copying a file on the Dropbox itself. In order to copy the file to your local machine you need to use the "get" argument.

./dbxcli get /susie_wedding/11-2009-wedding1.jpg

The Dropbox command line tool can do many other things as well but the ability to list Dropbox files and to retrieve a file to the local machine is enough for this project.

Weather

There are undoubtedly a lot of different options for getting the weather for a given city. If you do not need commercial quantity of weather there is a free solution from openweathermap.org. They provide both free and paid services based on the number of calls per month as well the number of calls per minute.

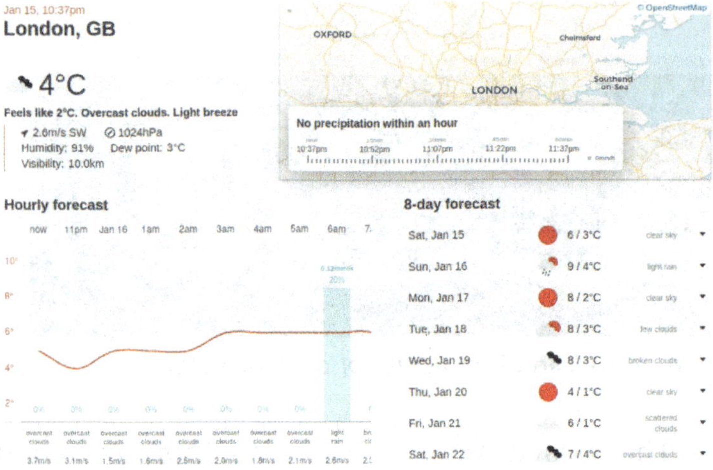

Figure 19: Sample of what is available from Openweathermap.org

It is rather amazing to see the types of services that are provided.

- Current weather data
- Hourly forecast 4 days
- Daily forecast 16 days
- Climatic Forecast 30 days
- Solar Radiation
- Global Weather Alerts
- 5 day 3 hour Forecast
- Historical Weather data
- Weather Maps
- Global Precipitation

- Air Pollution
- Geocoding
- UV Index

The historical data is also available for most of the types of "actual" data that they support.

Once your account has been created you can create one or more application keys which is then required for retrieving weather information. This allows you, Figure 20, to create different keys for different weather applications.

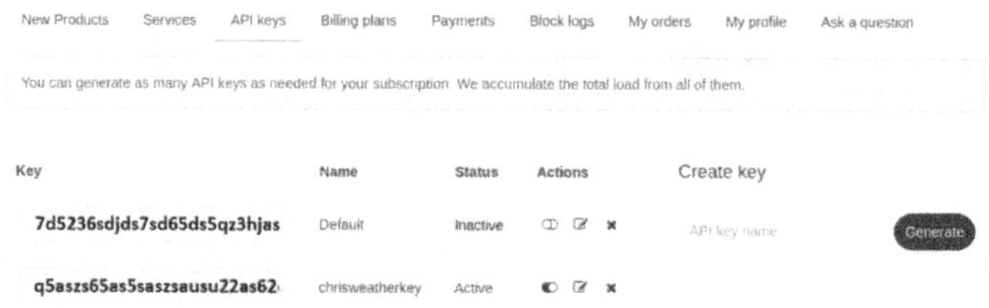

Figure 20: Managing keys in Openweathermap.org

It is possible to disable or delete keys as your application landscape changes. This is a cool feature for a company but perhaps a bit overkill for the average user. Even a free account can make up to 1,000,000 calls per month with a few exceptions, see the pricing page of their web site for more information.

Current weather

Openweathermap.org has a number of different types of calls that can be used to gather weather data. This can be done by using the city name or zip code as a filter. With some experimentation you can eventually get the weather for your area but there other filters.

There exists a list of locations each having their own integer code. This helps to distinguish between Fairview, New Jersey and Fairview, New York.

http://bulk.openweathermap.org/sample/

A free service to retrieve the weather is good but a free service that delivers its data with a simple internet call (ie REST API) is invaluable.

API call
api.openweathermap.org/data/2.5/weather?id={city id}&appid={API key}

One of the truly helpful things about the Linux environment is some of the supporting commands. Curl[16] is a command line tool that is used for transferring data to and from a server. It can be also be used to make a specific call to retrieve the weather.

Example
curl -s "http://api.openweathermap.org/data/2.5/forecast?id=5116819&appid={API_KEY}&units=imperial"

The data is returned as a JSON structure by default[17]. This data can be parsed programmatically which is good but at first

16 Client for URL

17 It is also possible to have the data returned as XML

glance the data is a bit overwhelming to look at, Listing 1.

Note: The curl command needs to be surrounded by quotes because of the use of ampersands(&) that are part of the argument.

Returned data

*{"cod":"200","message":0,"cnt":40,"list":
[{"dt":1642291200,"main":{"temp":8.51,"feels_like":-
1.86,"temp_min":8.51,"temp_max":10.81,"pressure":1028,
"sea_level":1028,"grnd_level":1019,"humidity":48,"temp_k
f":-1.28},"weather":
[{"id":804,"main":"Clouds","description":"overcast
clouds","icon":"04n"}],"clouds":{"all":100},"wind":
{"speed":6.11,"deg":32,"gust":15.39},"visibility":10000,"p
op":0,"sys":{"pod":"n"},"dt_txt":"2022-01-16 00:00:00"},
{"dt":1642302000,"main":{"temp":8.4,"feels_like":-
1.21,"temp_min":8.19,"temp_max":8.4,"pressure":1029,"se
a_level":1029,"grnd_level":1020,"humidity":50,"temp_kf":0
.12},"weather":*

`Listing 1 Unformatted JSON data`

Displaying data

The main task of the picture frame is to display the photographs which should be easy enough. Perhaps this could be done by setting up a screensaver to cycle through the files in a folder.

There is undoubtedly a number of different programs which could be used to display the pictures but there is a larger difficulty that needs to be overcome. Weather data could include some photographs but temperature forecasts tends to be text data. This complicates the task of displaying the textual weather data and photographs.

An alternative way would be to

programmatically create an image of the weather forecast.

Apache Formatting Objects Processor

Given enough time we could gather some libraries and write the software to create my weather data as a graphic image. Yet, rather than fall under the "not invented here" mentality I plan on using the Apache Formatting Objects Processor (FOP) to take my text data and convert it into a different format.

"Apache™ Formatting Objects Processor (FOP) is a print formatter driven by XSL formatting objects (XSL-FO) and an output independent formatter. It is a Java application that reads a formatting object (FO) tree and renders the resulting pages to a specified output."[18] Apache FOP can convert xml data into any of the following formats:

- PDF
- ASCII
- Text
- PostScript
- PCL
- RTF
- PNG
- TIFF

The most common use of this tool might be to convert XML data into PDF files. The strength of Apache FOP is that you take the same input files and create any of the supported output formats from the same input

18 https://xmlgraphics.apache.org/fop/

source.

It is possible to not only use text but also include graphics in the final output, so this program can be used to convert weather data into pictures of weather data.

Installing the Apache FOP depends on which operating system you are using. Quite a few Linux distributions have this software available in their package manager. This is also true for the Raspberry Pi.

```
sudo apt-get install fop
```

Displaying images

The most important program for digital picture frame is the ability to display photographs. One program which can do this is "feh" - a fast and light image viewer.

This program, unlike others, does not have a graphic interface for interacting with it. Although this might be a setback for casual users, it is perfect for using in scripts to display pictures. The pictures can be displayed in windows or can be full screen. There are even options for hiding the cursor and hiding the border.

options	
-y	hide pointer
-x	hide border
--scale-down	scale image down to fit screen
--auto-zoom	scale image up to fit screen
-F	full screen

The --scale-down option will ensure that photographs that are too large will be scaled down to fit and the –auto-zoom will make the pictures scaled up to fit the screen.

This program exists in the Raspberry Pi repository and can be installed with a single command.

```
sudo apt-get install feh
```

5 A BIG SCREEN

A digital picture frame needs a screen to display the photographs. The quick and easy solution would be to purchase a screen. There are a number of these screens ranging in size from 5" to 10" available for purchase over the internet. These screens are actually a premium solution as they also tend to support touch input. This functionality would not be used in this application but could be interesting for other development projects.

The cost of those screens range from 50 - 100 dollars. This cost is not exorbitant but with this price point it is possible to purchase a full sized monitor for slightly more than this amount. More important than this extra functionality is the difficulty to find a particular screen that is larger than 10" but smaller than 21".

15" and 17" screens

One source of this size screen would be to salvage one from an old laptop. There is nothing magic about the screen from a laptop when compared with a normal monitor. One small difference is that a monitor has a circuit board with the VGA or HDMI connector. The laptop connects the screen directly with the motherboard of the laptop and no connector is needed, but this is correctable.

Screen

Carefully disassemble the laptop extract the screen. In my case the screen has two rails, one on each side, which held the screen in place and gave it strength. In Figure 21 it is possible to see that the two rails of the frame that extend past screen to the top and bottom.

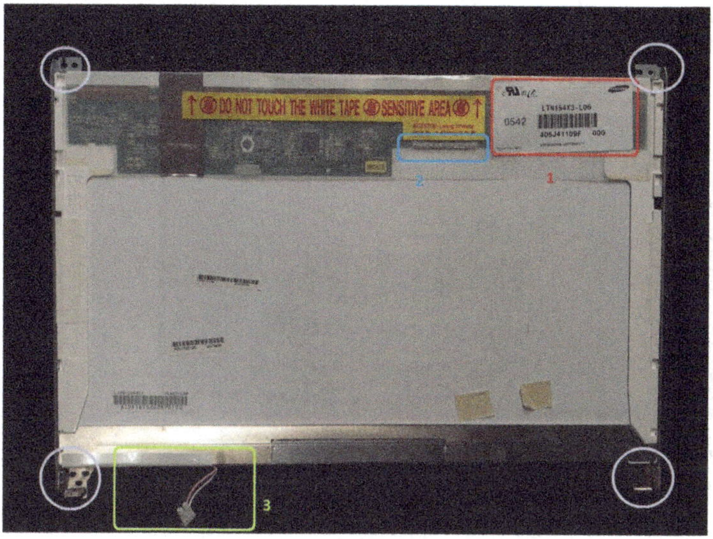

Figure 21: Back side of laptop screen

It is difficult to see in the picture but the blue circle shows where the controller board connects to the screen. This is where the video signal is fed to the screen, but it is the second connector, green circle, which controls the back lighting. The back lighting is simply some white leds which shine through and cause the picture to be displayed. In order to use a laptop screen as a standard screen a special controller board is required.

Unfortunately there is no universal controller board that can be used with all screens. Each screen requires its own specific controller board that correctly displays the images based on the physical characteristics. Typically each laptop screen will have a sticker with its own model number, Figure 22, so it is possible to find and purchase a compatible controller board.

Figure 22: Controller board, LTN154X3-L06, for this screen

All of these special controllers are much much less common than most PC components (ie CPU, Fan, Video card) and most likely will not exist in your local electronic shop nor on e-commerce giant Amazon. These types of controller do however tend to show up on Ebay or Alibaba.

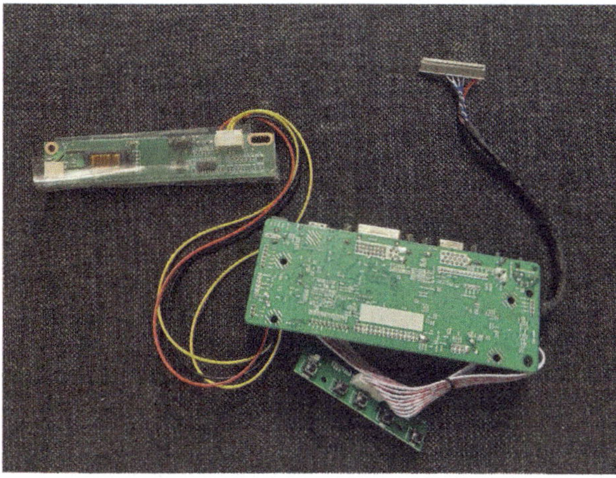

Figure 23: Video Controller Card

I managed to find the exact video controller board for my laptop screen on Ebay[19] for 20€. This controller card also includes a small PCB that has five buttons which can be used to control the output. This board is analogous to the menu and option buttons that are found on full size monitors.

This is a pretty good controller card as it outputs VGA, DVI and HDMI. Unfortunately this controller board didn't come with a power supply. The Ebay sale did mention that it requires 12v dc power supply.

Controller board specification

M.NT68676.2 LCD Controller Board

HDMI DVI VGA AUDIO LCD Board Work for LVDS Interface LCD Screen

Input Power Adapter:12Vdc More than 2A.

Come In Standby Model When Not Input Signal

19 Approximately 22 USD.

Standby Model <1W
Video Input: HDMI DVI VGA
Audio Input: HDMI
Audio Output: Speaker Connector
OSD: Brightness Contrast Auto Language ...
Language: English French German Spanish Italian ...
Board Size:125mm x 58mm x 17mm

This should be enough information but there is one detail that was not listed, the power connector. Mine is a barrel jack 5.5mm x 2.1mm but it is not clear if this is the same for all similar LCD controller boards.

Creating the Raspberry Pi screen is just a matter of connecting the video controller board and power to the laptop screen.

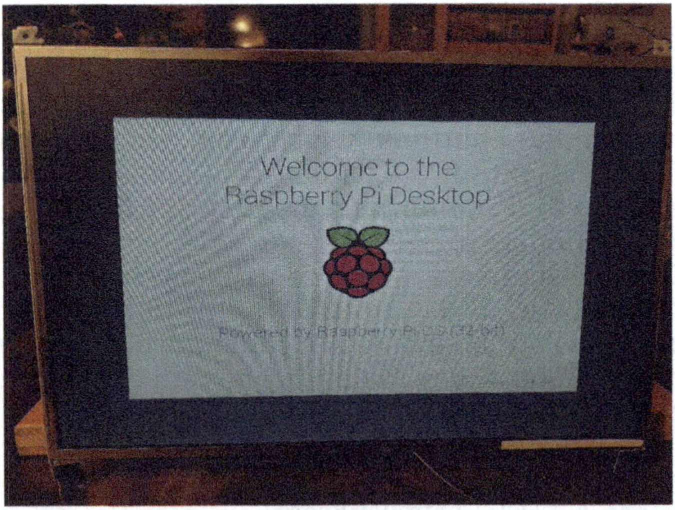

Figure 24: Laptop screen connected to a Raspberry Pi

This DIY screen can then be connected to

the Raspberry Pi or any other computer just like a normal monitor.

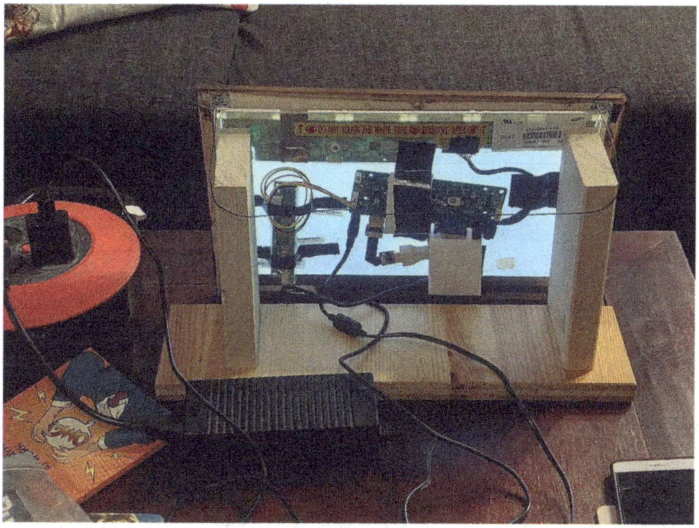

Figure 25: Temporary structure to hold screen up during development

6 PROGRAMMING THE FRAME

Before creating our scripts to control the smart frame it is important that all required software is installed. Most of these programs can be installed from the Raspberry Pi package repository if they are not already in the base operating system image.

Required software.

- Java
- curl
- jq
- feh
- FOP
- dbxcli

Java and curl are already installed with the full version Raspberry Pi OS installed on our SD card. Most of the programs can be installed from the command line.

sudo apt-get install jq feh fop

This command, much like what we saw on page 21, will display how much space will be required to install this software and then ask for approval to perform the install.

However, not all required software is available in the package repository. One such exception is the Dropbox client (dbxcli). This client can be downloaded with just a few commands.

1. mkdir /home/pi/bin
2. cd /home/pi/bin
3. wget https://github.com/dropbox/dbxcli/releases/download/v3.0.0/dbxcli-linux-arm
4. mv dbxcli-linux-arm dbxcli
5. chmod 777 dbxcli

This set of steps will download the this software into a specific directory and change the executable permissions for it. The location, /home/<user>/bin was not a casual location choice. The cool thing about the Raspberry Pi OS is that it was considered that the user may wish to run their own utilities, thus the /home/<user>/bin directory will automatically be added to the search path of the terminal if the directory exists.

Current weather

Their are quite a few different API calls for retrieving weather data. The picture frame will display both the current weather information as well as a forecast information. In order to display all of this information several calls may be necessary.

Current weather API by City Id

The current weather information that we require is relatively small. You will receive the local weather including temperature, sunrise, sunset, clouds as well as wind.

API call

```
api.openweathermap.org/data/2.5/weather?id={city
id}&appid={API key}
```

Here is a sample of the data that the above call would be return.

{"coord":{"lon":-93.3872,"lat":45.1977},"weather":[{"id":802,"main":"Clouds","description":"scattered clouds","icon":"03d"}],"base":"stations","main":{"temp":259.67,"feels_like":254.57,"temp_min":257.73,"temp_max":261.53,"pressure":1018,"humidity":51},"visibility":10000,"wind":{"speed":2.24,"deg":5,"gust":6.71},"clouds":{"all":29},"dt":1645131485,"sys":{"type":2,"id":2001347,"country":"US","sunrise":1645103509,"sunset":1645141414},"timezone":-21600,"id":5016450,"name":"East New York","cod":200}

Listing 2 Unformatted JSON data

All the information we need is in this JSON structure but this can be difficult to extract even using regular expressions. However, JSON structured data is the desired format for web most pages. Not all web data is processed using HTML and javascript so other developers have created tools to ease the task of processing this type of data structure.

The jq program, a command line JSON processor, is super for parsing out values from a JSON object. It can also be used to reformat JSON data. The above JSON block can be formatted into something a bit easier to read using jq.

```
ie.
     jq "." inputfile.json
```

This command will convert the JSON data, Listing 2, to this formatted output.


```
{
  "coord": {
    "lon": -93.3872,
    "lat": 45.1977
  },
  "weather": [
    {
      "id": 802,
      "main": "Clouds",
      "description": "scattered clouds",
      "icon": "03d"
    }
  ],
  "base": "stations",
  "main": {
    "temp": 259.67,
    "feels_like": 254.57,
    "temp_min": 257.73,
    "temp_max": 261.53,
    "pressure": 1018,
    "humidity": 51
  },
  "visibility": 10000,
  "wind": {
    "speed": 2.24,
    "deg": 5,
    "gust": 6.71
```

```
  },
  "clouds": {
    "all": 29
  },
  "dt": 1645131485,
  "sys": {
    "type": 2,
    "id": 2001347,
    "country": "US",
    "sunrise": 1645103509,
    "sunset": 1645141414
  },
  "timezone": -21600,
  "id": 5016450,
  "name": "East New York",
  "cod": 200
}
```


Listing 3 Formatted JSON data

If you are limited to using shell scripts and do not need too much data you can easily use jq for processing this data. To use jq, the first step is to fetch the values and store them in a variable.

```
APPID=323737373ee29a6832373237373737374c
CITYID=5115985   # east new york
TMP=/var/tmp
MEASUREMENT=imperial
OFFSET=7
MAXFORECAST=7

geticons()
{
  icon=$1
  if [ ! -f ${icon}.png ]
  then
    echo fetching graphics for $icon
    wget http://openweathermap.org/img/wn/$
{icon}.png
    wget http://openweathermap.org/img/wn/$
{icon}@2x.png
```

```
    wget http://openweathermap.org/img/wn/$
{icon}@4x.png
  fi
}

getcurrentweather()
{
  CMD="api.openweathermap.org/data/2.5/weather?
id=${CITYID}&appid=${APPID}&units=$MEASUREMENT"
  RET=`curl -s $CMD`
}
```

Listing 4 Bash script with getcurrentweather function to retrieve current weather

The getcurrentweather function will make a call to openweathermap.org and the result from the call will be assigned to the variable RET. The variable RET will contain the above JSON so further values can be extracted later.

```
setcurrentvalues()
{
  weatherdesc=`echo $RET | jq -r
".weather[0].description"`
  currenttemp=`echo $RET | jq -r ".main.temp"`
  expectedlow=`echo $RET | jq -r ".main.temp_min"`
  expectedhigh=`echo $RET | jq -r
".main.temp_max"`
  cityname=`echo $RET | jq -r ".name"`
  currenticon=`echo $RET | jq -r
".weather[0].icon"`

  sunriseDT=`echo $RET | jq -r ".sys.sunrise"`
  sunriseHH=`date --date=@$sunriseDT +'%H'`
  sunriseMM=`date --date=@$sunriseDT +'%M'`
  sunriseHH=$(($sunriseHH - $OFFSET))
  sunrise=$sunriseHH:$sunriseMM

  sunsetDT=`echo $RET | jq -r ".sys.sunset"`
  sunsetHH=`date --date=@$sunsetDT +'%H'`
  sunsetMM=`date --date=@$sunsetDT +'%M'`
  sunsetHH=$(($sunsetHH - $OFFSET))
```

```
  sunset=$sunsetHH:$sunsetMM

  currenttime=`echo $RET | jq -r ".dt"`
  currenttime=`date --date=@$currenttime +'%Y-%m-
%d'`

  geticons $currenticon
}
```

Listing 5 Bash script with setcurrentvalues function to extract values from JSON object

The setcurrentvalues function extracts the various values from the RET variable. Selecting a value from the JSON structure is pretty simple.

One part of the JSON data that is not very intuitive is why the dates are very large numbers. These are actually the number of seconds since Jan 1, 1970. The Linux date command can interpret and reformat these numbers into user friendly values. It is because of this conversion you see the date command being used in the script.

There is another function that is called geticons, Listing 4, has not yet been described. This function will download a tiny icon that matches the weather description. The weather forecast will return a number such as 04d or 04n. This icon is available in three different sizes - normal, double size, and quadruple size. These graphics will be used later when we create the weather forecast graphical image. The geticons function will fetch only the icons that are currently necessary, but over time the collection will grow as the weather conditions change.

Forecast by City Id

It is possible to retrieve forecast values for up to 5 days.

API call

```
api.openweathermap.org/data/2.5/forecast?id={city ID}&appid={API key}&cnt={count of desired}
```

The forecast call will return an JSON array of forecast values with the changing conditions every three hours for up to five days. Despite how large an individual forecast appears to be, most of the data is not terribly interesting for most people.


```
{
  "cod": "200",
  "message": 0,
  "cnt": 1,
  "list": [
    {
      "dt": 1643587200,
      "main": {
        "temp": 22.64,
        "feels_like": 13.37,
        "temp_min": 20.19,
        "temp_max": 22.64,
        "pressure": 1018,
        "sea_level": 1018,
        "grnd_level": 1020,
        "humidity": 60,
        "temp_kf": 1.36
      },
      "weather": [
        {
          "id": 802,
          "main": "Clouds",
          "description": "scattered clouds",
          "icon": "03n"
        }
      ],
      "clouds": {
```

```
        "all": 34
      },
      "wind": {
        "speed": 8.12,
        "deg": 257,
        "gust": 12.55
      },
      "visibility": 10000,
      "pop": 0,
      "sys": {
        "pod": "n"
      },
      "dt_txt": "2022-01-31 00:00:00"
    }
  ],
  "city": {
    "id": 5115985,
    "name": "East New York",
    "coord": {
      "lat": 40.6668,
      "lon": -73.8824
    },
    "country": "US",
    "population": 0,
    "timezone": -18000,
    "sunrise": 1643544438,
    "sunset": 1643580612
  }
}
```


Listing 6 Sample forecast data

This function getforecast call only returns a single forecast (ie &cnt=1) but the retrieval function can just as easily download twenty forecasts.

```
getforecast()
{
    CMD="api.openweathermap.org/data/2.5/forecast?
id=${CITYID}&appid=$
{APPID}&units=$MEASUREMENT&cnt=$MAXFORECAST"
  RET=`curl -s $CMD`
}
```

```
setforecastvalues()
{
  IDX=$1
    forecasttemp=`echo  $RET  |  jq  -r  ".list[$
{IDX}].main.temp"`
    forecastdate=`echo  $RET  |  jq  -r  ".list[$
{IDX}].dt"`
   forecasttime=`date  --date=@$forecastdate  +'%H-
%M'`
   forecastdate=`date  --date=@$forecastdate  +'%Y-
%m-%d'`

    forecastdesc=`echo  $RET  |  jq  -r  ".list[$
{IDX}].weather[0].description"`
    forecasticon=`echo  $RET  |  jq  -r  ".list[$
{IDX}].weather[0].icon"`
    forecastwind=`echo  $RET  |  jq  -r  ".list[$
{IDX}].wind.speed"`
   forecastpressure=`echo $RET | jq -r ".list[$
{IDX}].main.pressure"`
   forecasthumidity=`echo $RET | jq -r ".list[$
{IDX}].main.humidity"`

  geticons $forecasticon
}
```

Listing 7. Bash script with setforecastvalues and getforecast methods

Making Images

It is possible to create a PDF or a PNG file using the Apache Formatting Objects Processor. How the process works is that you create a template using xsl which contains all the formatting. How to define such a template is a very involved process.

Learning about how to create such a template can be found on the internet but it is also possible to purchase a book on the topic[20].

20 XSL-FO Tutorials - Herong's Tutorial Examples, ISBN 979-

It would be nice to completely explain this as well but it could take dozens of pages to give an incomplete description. The code for the template, as well as the rest of the project is available for download from its github repository.

`https://github.com/dockraspberrypi/smartframe`

Similar to writing a computer program there are many different ways to achieve a solution using FOP. One of the easiest ways to use xsl-fo is to structure your data that it fits to the template flow. It is convenient to structure the data so it appears in the file in the same order as it needs to be filled into the form.

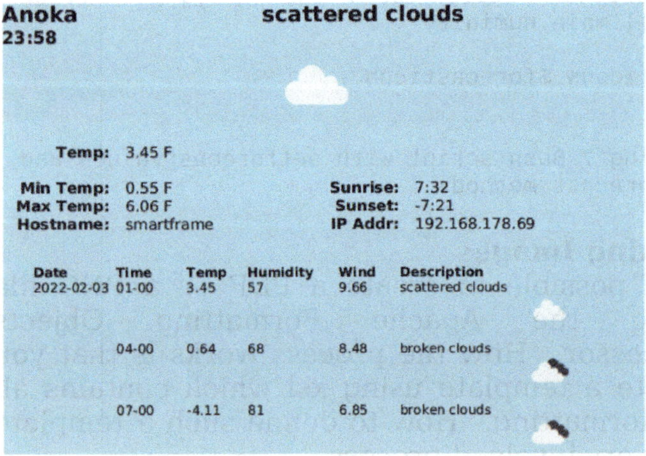

Figure 26: Example output from Apache FOP

In this example, Figure 26, we can visualize

the data into three different groups.

- Header information (city, desc, time)
- Temperature info(min,max,sunrise,...)
- Forecast info (date,time,humidity,...)

The code for creating the xml data from a shell script is not really difficult. Simply echo the data and redirect it to a specific file.

ie.
echo "<header>" > myfile.xml

Rather than showing a few dozen lines that do just this, perhaps it is more instructive to look at the structure of the xml created.

```
<?xml version="1.0"?>
<weather>
  <header>
    <todaydate>2022-02-01</todaydate>
    <nowtime>23:41</nowtime>
    <cityname>Anoka</cityname>
    <weatherdesc>scattered clouds</weatherdesc>
    <weatherpic>03d@2x.png</weatherpic>
  </header>
  <current>
    <temp>12.45 F</temp>
  </current>
  <minline>
    <mintemp>8.91 F</mintemp>
    <sunrise>7:33</sunrise>
  </minline>
  <maxline>
    <maxtemp>16.03 F</maxtemp>
    <sunset>-7:20</sunset>
  </maxline>
  <hostline>
    <hostname>smartframe</hostname>
    <ipaddr>192.168.178.69</ipaddr>
  </hostline>
```

```
  <forecast>
     <single>
        <date>2022-02-02</date>
        <time>01-00</time>
        <temp>12.33</temp>
        <pressure>1014</pressure>
        <humidity>66</humidity>
        <wind>15.5</wind>
        <desc>scattered clouds</desc>
        <icon>03n.png</icon>
     </single>
       …
  </forecast>
</weather>
```

Listing 8 XML data used to create weather image

The different sections of this xml code have been highlighted. The header section, highlighted in green, contains the city name, weather description, current time and the name of the picture for this description.

The next four highlighted sections (blue, yellow, peach and orange), each represents a single line on the weather image. One example, yellow, is minimum temperature and the sunrise.

The template and xml was designed so that as the data is encountered in the data file it is then output to the picture.

The final section(s), purple, represents a single forecast line. The template is driven by the data and there is no specific limit to how many forecasts will be processed. Because of this, the template will continue to output the forecasts lines while data exists in the XML file.

Downloading pictures

Downloading the photos from Dropbox is not complicated but it helps to know the structure of how the photos are stored. Do all files exist in the root directory? Are they stored in one or more subdirectories. I have chosen to organize all my photographs in separate directories.

```
pi@smartframe:/var/tmp $ dbxcli ls -l
Revision Size Last modified Path
-        -    -             /Halloween2013
-        -    -             /2020-2021
-        -    -             /susie_wedding
pi@smartframe:/var/tmp $
```

Figure 27: Listing of directories on Dropbox

With the data stored in this way it is possible to gather up the list of subdirectories with a single, although somewhat complicated command.

```
pi@smartframe:/var/tmp $ dbxcli ls -l | grep -v Revision | cut -c 29-
/Halloween2013
/2020-2021
/susie_wedding
pi@smartframe:/var/tmp $
```

Figure 28: Command to extract directory names only

The grep portion of the command eliminates the titles and the cut command will eliminate the spaces and dashes. When run it will return a list of directories that are ready to be used to gather their own file listing.

This list of directories can be assigned to an environment variable and then processed one at a time.

```
pi@smartframe:/var/tmp $ DIRS=`dbxcli ls -l | grep -v Revision |cut -c 29-`
for i in $DIRS ; do echo $i; done
/Halloween2013
/2020-2021
/susie_wedding
pi@smartframe:/var/tmp $
```

Figure 29: Assign directories to environment variable

It is then possible to do a listing of any of those directories.

```
pi@smartframe:/var/tmp $ dbxcli ls -l /Halloween2013 | cut -c 33- | grep -v Path
/Halloween2013/10-2013-Halloween.jpg
/Halloween2013/10-2013-Halloween1.jpg
/Halloween2013/10-2013-Halloween2.jpg
/Halloween2013/10-2013-Halloween3.jpg
/Halloween2013/10-2013-Halloween4.jpg
/Halloween2013/10-2013-Halloween5.jpg
/Halloween2013/10-2013-Halloween6.jpg
/Halloween2013/10-2013-Halloween7.jpg
pi@smartframe:/var/tmp $ dbxcli get /Halloween2013/10-2013-Halloween.jpg
Downloading 6.6 MiB/6.6 MiB
pi@smartframe:/var/tmp $
```

Figure 30: Listing of files in a Dropbox directory

One of the slightly inconvenient things about using this Dropbox client tool is that although you can get and put datafiles you do need to use the entire qualified path when doing so.

The Raspberry Pi Zero 2 W is a small computer. It is possible to store quite a few pictures on the SD card especially if it is 16GB or 32GB in size. However, because the SD card does not have wear leveling it might be a bad idea to constantly download photos to it.

To get around this, the idea is to download a random picture every few minutes but instead of storing on the SD card it will be stored on the ram disk. The ram disk is actually quite small at 15MB so this will not be a long term storage as much as a temporary location prior to displaying a photo.

In order to randomly select a file we will need to retrieve a list of all files. Then a random file will be selected, downloaded and saved to the name "picture.jpg".

```
# collect a list of subdirectories
DIRS=`dbxcli  ls  -l  |  grep  -v  Path  |  sed
's/^.*\///g'`
for single in $DIRS
do
  # gather files for one of the directories
  FILES=`dbxcli ls -l $single | grep -v \
        Path | sed 's/^.* \///g'`

  # collect all files together
  TOTALFILES=`echo $TOTALFILES $FILES`
done

# convert files into an array
myArray=($TOTALFILES)

# total items in the array
cnt=${#myArray[@]}

# calculate a random # between 0 - count
idx=$(( $RANDOM % $cnt ))

dbxcli get ${myArray[$idx]} picture.jpg
```

Listing 9 Bash script to download a random picture

The last few lines of code will get a list of all files and copy a random file locally. The feh program will display a picture on top desktop which is what we want, however, if we stop the current picture before displaying another one there will be a brief display of the desktop. This is a bit annoying, so it is important to display the new picture before stopping the old one. Stopping the display of an old picture can only be done by killing the feh process displaying the picture.

In order to kill a running feh process it is necessary to retrieve the process id. This sample, Listing 10, will retrieve the id of the current picture, if it is running, and will display the new file. Once the new picture is displayed the old one will then be removed.

```
export DISPLAY=:0.0

# get id of current picture
CURRENTPIC=`ps -ef| grep feh | grep -v grep | awk
'{print $2}'`

# display a new pic
feh -y -x --scale-down --auto-zoom bob.jpg &

# now kill old pic
sleep 1
if [ ! -z $CURRENTPIC ]
then
  echo kill $CURRENTPIC
  kill $CURRENTPIC
else
  echo not set
fi
```

Listing 10 Bash script to display a picture

Scheduling

There is nothing preventing us from running a single big script in a loop with a few delays between the various actions.

A more elegant method would be to use the built in cron scheduler. With cron, we can schedule various tasks by the minute, hour, day of month, or day of week. This will make it easy display the pictures during normal times people are awake such as 6am to midnight. The pictures could be displayed twice an hour and we can also display the weather periodically.

```
6,36 6-23 * * * /home/pi/smartframe/display.sh
0,30 6-23 * * * /home/pi/smartframe/weather.sh
```

Listing 11 Crontab listings

These two entries, Listing 11, show how this could be possible; we could display the weather at the top and bottom of the hour and then six minutes later we that picture will be taken down and a new photo would be displayed. This would take place between 6am and midnight.

7 CARPENTRY

Professionally framed photographs have a lot of parts - picture, matting, glass and wooden or metal frame. This looks great for photographs but most of this is not really necessary for a digital picture frame.

Figure 31: Screen from laptop

One of our goals is to transform the laptop

screen into a digital picture frame but more importantly, it should not remind us of a piece of old electronics.

I am not a professional woodworker so I went to the lumberyard to get some ideas. In the end I purchased some baseboard molding and some other small pieces of wood which could be fashioned into a thin box to hold the electronics on the back of the screen.

No specials tools are necessary but like any other situation the more special tools you have the easier it can be to perform some tasks. If you have a mitre box it is easier to get proper 45 degree angle cuts so the frame fits together without any gaps.

Figure 32: Mitre box

Yet, creating a picture frame is really just the act of creating a small flat box with a couple of specially formed pieces of wood.

I did consider purchasing a frame on the internet. This would have given me the option of choosing which type of wood, what color it should be painted or stained as well as a piece of glass. A piece of glass seemed unnecessary and might even cause more reflection making the viewing difficult. I am certain these custom made frames would have been lovely but they were also rather pricey excluding the shipping which was fairly substantial as well.

Figure 33: Self made frame

Creating the box for the back of the screen is considerably easier due to its hidden nature. Tiny imperfections will not attract the eye so not all cuts need to be perfect.

Figure 34: Box to hold the electronics

The steps are pretty simple.

- Pick out some wood for frame
- Measure and cut frame
- Measure and cut wood for box
- Glue frame together
- Glue box together
- Stain frame

There is very little that needs to be described to create the frame. Building the box can be as complicated as the carpenter wants. I was lucky that it was possible to purchase some wood that was L shaped. I was able to use this wood to create the box for the back of the frame.

The pieces of wood were cut with a 45 degree ends to be assembled in a similar manner to the frame. In order to add some additional strength to the wooden box I added a few pieces of scrap wood in the corners. I used some stain so the frame wouldn't be raw unfinished wood.

8 FINAL ASSEMBLY

One of the luckiest decisions was to keep the supporting frame attached to the screen. These side frames add additional strength so the screen doesn't flex. They also provided a solid point to attach the frame and box to.

The frame has small tabs at the top of the screen and at the bottom where it previously connected to the hinges. The tab that attaches to the hinge was slightly tilted which made it more challenging to use.

The frame

One complication for my digital frame when compared to more traditional picture frame is that it is relatively thick. This prevented me from having simple a hook on the back of the frame. The solution was to add a cable to hang the frame.

Figure 35: Laptop screen with glue on support tabs

It was not clear where I could find a small thin cable so I took three pieces of small diameter wire and wound them together. The result was both very robust and looked quite nice.

The real task was to attach the frame to the screen. I wanted to do this without using too much glue in case the screen needed to be removed for upgrades. To do this I added glue to the tabs. When the frame was set on top of the screen the glue only made contact to two of the four points. This was because the baseboard molding was not flat on the back. It was also because two of the tabs were tilted.

It was required to add even more glue to make contact[21]. When the glue was almost dried I added two additional points on the back side of the frame to ensure it doesn't get loose.

Box to hold electronics

It is not visible but when I assembled the frame, I glued small pieces of tin on the back side of the frame on the corners. This was to provide extra support.

I used some more tin to connect the box on the back to the screen frame.

  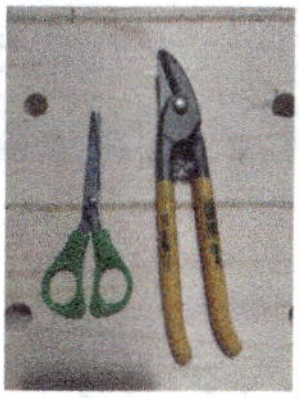 

I did not have enough tin, but I did have an aluminum can that I cut up and used to connect the box to the frame. In the end this was the best decision. The aluminum can be cut with a pair of scissors but is still strong enough for holding everything together.

The wooden box was attached to the back and had a strong connection because I used

21 Before adding more glue I needed to wait 45 minutes for it slightly harden before adding more. Otherwise too much glue would run.

four aluminum tabs to connect it to the frame with screws. I did use a small amount of glue to help ensure a strong connection.

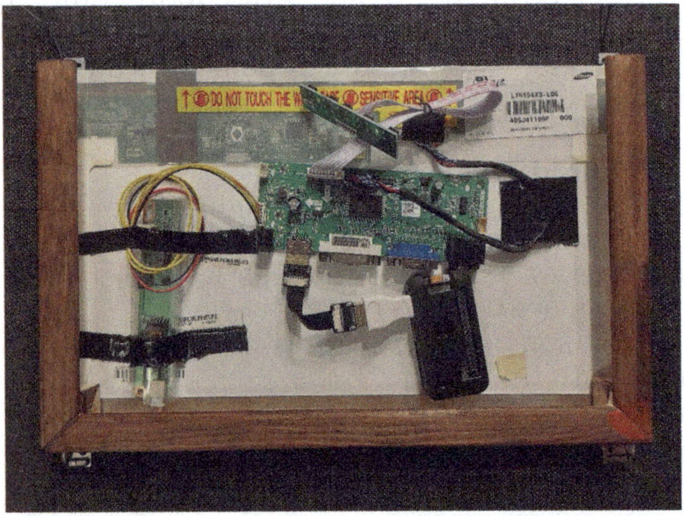

Figure 36: Frame with the parts taped to the back of the screen

Unexpected Modifications

The initial plan was to add a thin piece of wood backing to complete the box and ensure the parts would not fall out as well as providing some protection to the electronics.

During an unplanned visit to the lumberyard I spotted a piece of acrylic plexiglass in the bin of offcuts. It was larger than I needed but not by much. I thought it would be interesting to make the back of the box see through.

Figure 37: Plexiglass clamped down

There are a number of different methods for cutting acrylic plexiglass. It can be cut with a saw but the saw should be quite sharp. A second method is to score the plexiglass with a sharp knife in a straight line on both sides and then press on it. This second method will cause the fracture on this line.

There is an actual special tool for scoring this line, which I do not own, but I did own a box cutter with a new blade. I scored the line on both sides both sides of the plexiglass perhaps about 10 or 12 times. When I applied the pressure I did get a clean break for almost the entire line. Yet, "almost" is not quite the same as "all". There was a piece perhaps six cm long that still needed to be removed. The plexiglass was still clamped between two pieces of wood so I was able to make a second attempt to break this. I was able to use a pair of pliers to break off the rest. I drilled a few holes through the plexiglass so I could screw it down to the box.

The plexiglass looked so good and worked so well that I removed all of the duct tape, Figure 36, and used hot glue to attach all items to the back of the screen.

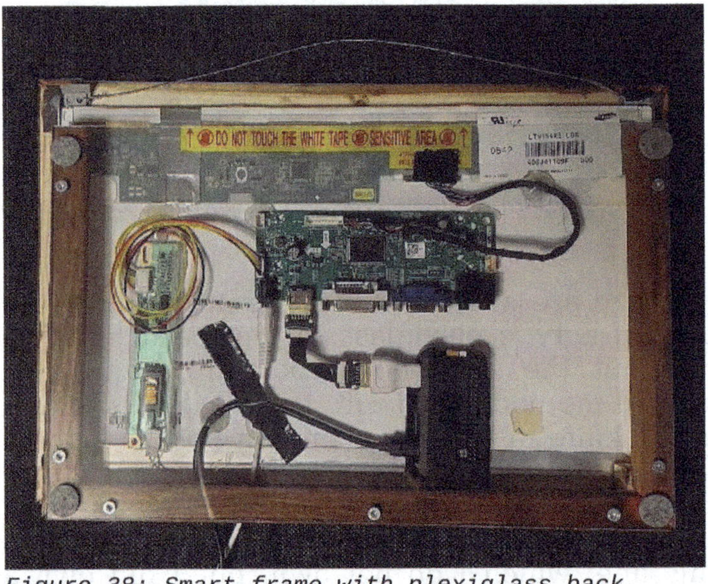

Figure 38: Smart frame with plexiglass back attached

9 TROUBLESHOOTING / LESSONS LEARNED

It would be neat if everything always went as planned, but things don't always work out that way in the real world. There were a few things that didn't work out initially but despite that the project still turned out ok.

Networking troubles
In the past when the Raspberry Pi was first released there were times that the wireless networking would become disconnected[22]. I do not know how frequent the problem is but this is a pretty serious issue for an internet connected device. The solution was actually an easy fix. Simply test to see if a known ip address is reachable, if not then restart the networking module. I wrote a small script, Listing 12, to do this task.

22 This did occur a few times during development

```
#!/bin/bash
ping -c 1 8.8.8.8
if [ $? -ne 0 ]
then
   echo restarting network >> /tmp/checknet.log
   echo `date `>>  /tmp/checknet.log
   /etc/init.d/networking restart
fi
```

Listing 12 Network test and restart script

In this case, the ip address used was 8.8.8.8 which is the Google DNS server. This script needs to be scheduled by a user with the necessary permissions to restart the networking module. The easiest way to do this was to schedule,Listing 13, this script via cron for the root user.

```
#
# m h  dom mon dow   command
* * * * * /home/pi/bin/checknet.sh
```

Listing 13 cronjob schedule of networking script

Acrylic backing

Acrylic plastic is clear like a pane of glass but has superior impact resistance and is lighter than glass too. It is more similar to plastic so it can be cut with a saw. Because of its composition you can score a line where you want it to be broken, clamp it and press down and it will fracture on that line. This is the theory, but the one thing not mentioned is that the acrylic is brittle.

Figure 39: Results from drilling too fast with a dull wood bit

Acrylic can bend a small amount but it will shatter if you put too much pressure on it. You can drill through it but if you are not careful the acrylic will fracture. The best advice is to use a sharp drill bit and drill slowly. I was lucky as I did some practice with a few left over scraps of the acrylic glass. My first attempt caused the acrylic scrap to fracture.

Use safety equipment

In this project I used both tin and small aluminum strips. I had an old pair of cutters which worked fine when cutting the tin but was quite awkward when cutting up the aluminum can. An aluminum can is really quite thin and can be easily cut with a normal pair of scissors.

I used a drill to make my holes in the aluminum but the gloves make the process clumsy and slow. I took the gloves off for the last set of holes and lacking the gloves the aluminum did spin and cut my finger.

Wood glue

I did use some aluminum strips to attach the box to the back of the screen but I also used some wood glue to affix the wooden frame to the screen frame. Wood glue is not intended to be poured in large pools but rather to be spread thinly between two pieces of wood and then clamped together. My smart frame is neither in regular dimensions nor strong enough to clamp.

The pea sized glue drops did managed to connect two parts of the frame but not the other two. It took days for this to dry. If you choose this assembly method ensure it has lot of time to dry. Otherwise it will not properly hold and the weight of the frame might cause it to separate.

Staining

I was fortunate when gluing the wooden frame together that no glue bled out onto the frame. I was less lucky with the box on the back of the screen. It appears that the glue was not properly cleaned off and once stain was applied it became a glaringly obvious, Figure 40.

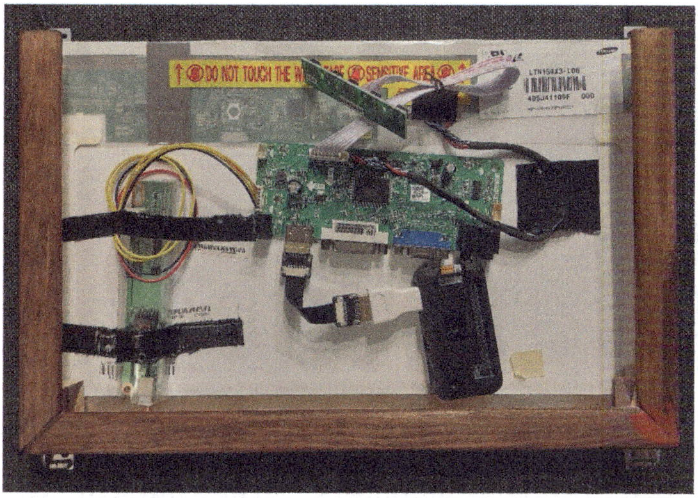

Figure 40: Glue on wood, lower right corner, caused odd stain spot

Creating picture frames

I used floor molding for my picture frame. It has a decorative shape on top but the bottom was not flat. This caused problems with trying to assemble as the cuts were close to 45 degrees but not exact.

Saving money

It is possible to buy a lot of accessories for the Raspberry Pi such as a case. This might make a very attractive solution, but if the Pi is not visible you can probably spare this part. The Pi is not made of cotton candy and works even when exposed to the elements, well, sitting on a shelf or in a drawer. The only reason I used a case for this project was because it was left over from previous project.

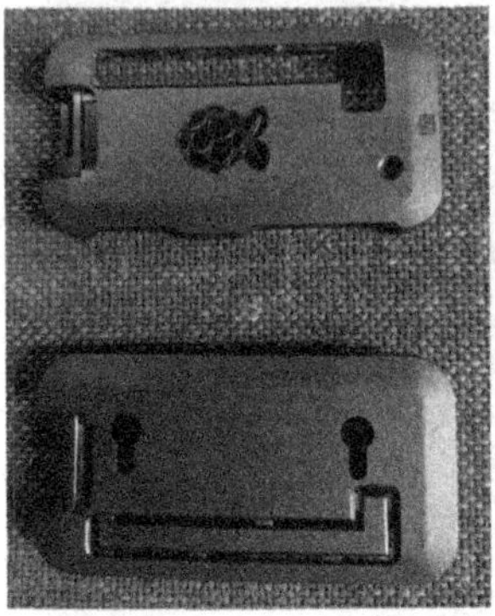

Proper development environment

The entire Raspberry Pi family, except the Pico, are all proper computers. They all run the same 32bit operating system and behave similarly.

The Raspberry Pi 4 with 4GB of ram is not as powerful as an Intel powered desktop computer but it is a very respectable computer and is easy to use. You can boot up into the desktop, surf the web, and do all of the development related to this project as well. The Raspberry Pi Zero 2 W is perfect as the brain for this project but a bit weak for development of this task. You could develop this project using a Pi Zero 2 W but it can be frustratingly slow at times.

If you prefer you could use any other desktop computer running Linux for development. Other Linux powered computers would be both similar to the Raspberry Pi and faster than the Pi Zero 2 W. A Linux PC is a great development platform but make sure that you use the correct dbxcli for that platform.

Timing

The only problem with displaying both the weather forecast and pictures using the crontab is that no photograph will be displayed initially on bootup. If you are lucky one minute later an image will be displayed but if you are unlucky you will see the desktop of the Raspberry Pi for quite a few minutes.

Fortunately there is a solution to this problem. It is possible to schedule a task in the crontab on reboot, Listing 14.

```
#!/bin/bash

export DISPLAY=:0.0
sleep 120
CNT=`ls -l /home/pi/smartframe | wc -l`
if [ $CNT -ge 1 ]
then
  FILE=`ls -1 /home/pi/smartframe/*jpeg | head -1 `
  /usr/bin/feh -F -x -Y --scale-down $FILE &
fi
```

Listing 14 Display image after boot

This small script will wait for 2 minutes and then display the a default picture. The two minute delay can be reduced after some testing.

```
#
# m h  dom mon dow   command
@reboot su pi -c "/home/pi/smartframe/initpicture.sh"
```

Listing 15 Schedule initial image script

This script, Listing 4, is scheduled in the root crontab, Listing 15.

Everything goes black

I reused a very old Raspberry Pi for this project and old SDCard. I shut it down and the next day I tried to start it it failed.

During my investigation I created a new SD card and reinstalled all of my scripts. I was surprised when after a short time the screen blacked out. Eventually I discovered that the image was configured with screen blanking enabled. This caused the screen to blank when there is no keyboard or mouse movement.

This feature can be turned off from the raspberry pi configuration program. This setting is located under display options.

10 APPENDIX - BASH

The Linux operating system and its many command line programs are different from Microsoft Windows yet many of the concepts are similar.

Bash - Bourne again shell

Computers and their windowing environments are intuitive and rather friendly but it was not always so. Before the windowing environments existed interfacing with the computer was done using the command line. This piece between the human and the computer was typically some form of interpreter. There have been many different flavors of shell interpreters over the years with the most widely used shell for Linux is the Bourne again shell or commonly referred to as bash shell.

The bash shell will execute system commands you type in but it is also used to interpret files full of commands. These scripts are typically a list of system commands, flow control statements or even calculations.

Like most languages it is possible to organize your code in separate functions to reduce complexity and enhance readability. This simple function, Listing 4, has a lot to unpack.

```
function areaofrectangle()
{
  width=$1
  height=$2
  echo $(($width * $height))
}
```

Listing 16 Function to calculate area

The bash shell was created in 1989 and when compared to newer languages it sometimes shows its age. Bash functions, just like the bash script itself, accesses the parameters in the order they are passed in. The parameters do not have names but rather just their position (ie $1). This function is assigning the incoming values to local variables, this is not required but for complex functions named variables makes for easier reading and comprehension. What might be less obvious is when doing assignments there can be no space between the variable, the assignment operator and the incoming variable (ie $1). Modern languages are more flexible with respect to white space.

Bash functions can return values but it is limited to integer values between 0 and 255. This can be done using the return keyword and the value.

```
return 23
```

The returned value will be accessible to the caller via the special variable “$?”. This example does not return its calculated value as a returned value. This is because it is not known in advance if the user will ask for a calculation that exceed this range for return values such as 20 x 20 or 100 x 3. The solution to get around this limitation is to perform an echo of the calculated value to standard output. This value can be captured and assigned to a value. It works to pass out values but is much less elegant then most modern languages.

The bash shell can perform integer calculations. These calculations are executed with the following syntax.

```
$(( something here ))

$(($width * $height))
```

Thus the final line of our areaofrectangle function performs the calculation and echos the result. An entire script to calculate the area of a rectangle might look like Listing 17.

```
#!/bin/bash

#
# calculation method
#
function areaofrectangle()
{
  width=$1
  height=$2
  echo $(($width * $height))
}

area=`areaofrectangle 2 3`
```

```
echo $area
```

Listing 17 Sample script calling calculate area

Listing 17 is virtually identical to Listing 16, but the differences are notable. The first line of the script tells the computer which interpreter should be used for this script. This line is important as there are many different shell interpreters that could be used.

```
area=`areaofrectangle 2 3`
```

The results of our function is being assigned to the variable "area". The function areaofrectangle is being called with a width of 2 and a height of 3. The backtic characters are not a mistake but rather tell the bash shell to evaluate what is between them and assign that value to our variable.

The Linux date command can be used for displaying the current date and time or to set the date and time.

```
date
> Sun 27 Feb 2022 03:36:07 PM CET
date +%Y-%m-%d
> 2022-02-27
```

More commonly you may wish to retrieve the date and use it. This could be for performing date or time sensitive operations or simply to save your calculations to a dated log file.

Virtually all computer languages have control statements. While some languages have more variations of control statements all have something similar to the while statement.

```
while [ some condition ]
do
        statement 1
        statement 2
        ...
done
```

Listing 18 Bash example while control statement

The if then control structure can also test conditions and make decisions based on them.

```
If [ some condition ]
then
        statement 1
        statement 2
else
        statement 1
fi
```

Listing 19 Script with if control statement

There are quite a few different conditions that can be tested for.

```
-f          file exists
-d          directory exists
-z          variable is null
-eq         equals
-ne         not equals
-gt         greater than
-lt         less than
!           not
```

These conditions can be used to test either environment variables or files/directories on the file system.

```
#!/bin/bash

VAR=1
while [ $VAR -lt 10 ]
do
  echo $VAR
  VAR=$(($VAR + 1))
done
```

Listing 20 Example while statement

```
#!/bin/bash

VAR=
if [ -z $VAR  ]
then
  echo not var exists
else
  echo var exists
fi
```

Listing 21 Example if/then statement

Eventually, calculating values will not be enough and it may be necessary to write them to a log or data file. The process of writing the data is fairly rudimentary. Simply echo something and use the greater than symbol (>) to redirect the output to a file.

```
echo “special text” > datafile.txt
```

There are actually two different streams for output, one for output with the second for errors.

```
echo “special text” 2> error.txt
```

This redirection can be used to output something specific to a file or can be used to redirect output or errors from commands being executed. These errors can be due to insufficient permissions or other application errors. These two streams are actually separate and can be redirected to different files.

```
some-program > output.log 2>error.log
```

There are circumstances where it is not important to separate errors from the output. In this case it is possible to merge both streams together and redirect the output to a single log file.

```
some-program > output.log 2>&1
```

It is easy to create log files using these techniques. In addition you can use the date command so the log files have date and time stamps in addition to the log information, Listing 22.

```
#!/bin/bash

LOGFILE=logfile-`date +%Y%m%d`

function logtext()
{
  LOGTEXT=$1
  LOGDATE=`date +%Y%m%d`
  echo $LOGDATE $LOGTEXT >> $LOGFILE
}

logtext "starting program"
logtext "area of square = $(( 3 * 4 ))"
logtext "ending program"
```

Listing 22 Example of data redirection

10 APPENDIX - LINUX COMMANDS

The bash shell is a small layer between the human and the operating system. It will interpret commands that it does know such as the change directory command (cd), but if you type a command that the interpreter does not know it will search your PATH attempt to find that program. If the program is found then it is executed. It is these other command line programs which makes bash scripts so powerful.

Wget - Non-interactive network downloader
The wget is a free utility for non-interactive download of files from the Web. It supports HTTP , HTTPS , and FTP protocols. The advantage of this is command is that it can be run from scripts to retrieve files.

Curl - client URL
This is a command line tool can be used to transfer data to and from the server. This tool is especially useful for calling REST API endpoints during development. The common

methods that are used are GET, POST, PUT, and DELETE. It is possible to pass in customized headers as well as JSON data to the endpoints.

Sed - stream editor
This command line tool is typically used to transform text into a different format. Sed is used to modify text from files or streams of data. Sed is more than just an editor it also has a rudimentary language to help in converting the text.

Awk - A general purpose scripting language
This command line tool can be used to scan through text line by line and perform operations. It is often used to parse an input line into separate fields, but it can also compare the input against patterns and perform actions on the matched input.

Grep - Pattern search tool
This command line tool is used to match lines in the input. It is usually used to filter through data or file listings.

Kill - stop a running program
The kill command is used to send a signal to a running program. There are quite a number different signals ranging from signal hangup to signal kill. Well written programs can accept these signals and behave in the appropriate manner such as shutting down in an organized manner.

Ssh - secure shell

The bash shell is a convenient way to interact with with a computer without a GUI. Sometimes it is necessary, or desirable, to connect to a computer remotely. The secure shell program will connect the terminal to a remote computer securely. Once connected you can type your commands as if you were logged into that computer locally.

Cron

The cron daemon is a scheduling program for running processes at scheduled times. This program will run programs or scripts as often as once a minute or it can also be run on a certain day of the week or on a given date and time.

Crontab

The crontab is the configuration for cron. Each user has their own crontab configuration.

ABOUT THE AUTHOR

Christopher was born in the United States in the Midwest and started his career as a software developer but apparently became restless and made the move to Europe. He has spent most of his time working in IT departments for banks and large corporations, usually as a consultant but occasionally as a software developer. His work brought him in contact with different people from different cultures but typically in a workplace environment.

It was during one of the IT projects he was lucky enough to meet an old time electrical engineer from the former Soviet Union. During the daytime we worked on software and in the evenings he infused me with the joy of hardware development.

Made in the USA
Monee, IL
03 November 2026

67380153R00056